DATE DUE		
MAY 1 8 2011		
FEB 08 2012	JAN 17 2013	

Light & Sound

CHRIS OXLADE

Heinemann Library
Des Plaines, Illinois

© 2000 Reed Educational & Professional Publishing
Published by Heinemann Library,
an imprint of Reed Educational & Professional Publishing,
1350 East Touhy Avenue, Suite 240 West
Des Plaines, IL 60018

Customer Service 888-454-2279

Designed by AMR
Illustrations by Art Construction
Printed in Hong Kong

04 03 02 01 00
10 9 8 7 6 5 4 3 2 1

Library of Congress Cataloging-in-Publication Data

Oxlade, Chris.
    Light & sound / Chris Oxlade.
       p. cm. – (Science topics)
    Includes bibliographical references and index.
    Summary: Explores various aspects of light and sound, including
reflection, the bending of light, optical instruments, color, sound
waves, the sense of hearing, and music.
    ISBN 1-57572-774-9 (lib. bdg.)
    1. Light Juvenile literature. 2. Sound Juvenile literature.
[1. Light. 2. Sound.] I. Title. II. Title: Light and sound.
III. Series.
QC360.095 1999
535--dc21   99-12884
                                              CIP

Acknowledgments
The Publishers would like to thank the following for permission to reproduce photographs:
Still Pictures/Gerry Gletter, p. 5 top; Science Photo Library/NASA, p. 5 bottom; NHPA, p. 7;
Science Photo Library/David Parker, p. 9; Kodak, p. 13; Science Photo Library/Freeman D. Miller,
p. 15; Science Photo Library/Bruce Iverson, p. 17; The Stock Market/Russell Munson, pp. 18, 27;
Still Pictures/Jany Sauvenet, p. 19; Science Photo Library/Earth Satellite Corporation,
p. 21; Redferns Music Library/Mick Hutson, p. 24; Still Pictures/Yves Lefevre, p. 25; Redferns
Music Library/John de Garis, p. 28; Redferns Music Library/David Redfern, p. 29.

Cover photograph reproduced with permission of Science Photo Library/Gorgon Garrado

Our thanks to Jane Taylor for her comments in the preparation of this book.

Every effort has been made to contact copyright holders of any material reproduced in this book.
Any omissions will be rectified in subsequent printings if notice is given to the Publisher.

Any words appearing in the text in bold, **like this**, are explained in the Glossary.

Contents

Light and Sound

Light and sound both play a large part in our lives. They allow us to see the world around us, to hear what is going on, and to enjoy sights and sounds such as paintings and music. Most animals have sensitive eyes and ears that detect light and sound. Eyes and ears are just as important to animals as humans because animals use them for hunting and for sensing danger.

Light and sound are both forms of **energy**, and they can be changed into other kinds of energy such as **electrical energy** and **mechanical energy**. Light and sound have another thing in common—you can think of both as **waves** that spread out from where they are made.

Light sources

Light is made by a change of energy. For example, in a fire, light is made from the **chemical energy** in the material that burns. In a light bulb, light is made from electrical energy. Light travels away from its source in straight lines called rays. You cannot see these rays unless they enter your eyes, but you can sometimes see rays of sunlight as they illuminate dust in the air.

During the day, most of the light we see comes from the sun. The amount of light that reaches Earth from the sun is enormous, but it is only a tiny fraction of the light that the sun gives out. The light is created from **nuclear energy** in the center of the sun. Its energy supports most of Earth's life and provides the energy that makes the world's weather happen.

Catching sunlight

Solar energy is energy captured from sunlight. Solar cells turn light directly into electricity. Solar panels turn light into heat energy and use it to heat water. The amount of energy even in bright sunlight is quite small, so large panels are needed to capture a useful quantity of energy. For example, a panel of solar cells of about one yard (one meter) square is needed to produce enough electricity to work a light bulb—and only if the sun is shining brightly! However, solar energy is very environmentally friendly.

Blocking the light

A shadow is made when an object blocks the path of light rays and forms a dark area on the side away from the light. If the light that is not blocked then hits a surface, you can see the dark outline of the object.

The earth itself forms a large shadow because it blocks sunlight. Each night, the part of the earth where you live moves through the shadow. During the day, shadows made by sunlight move as the earth spins and the sun appears to cross the sky.

Long shadows form on the landscape in the early morning and late evening when the sun is low in the sky. At midday in the summer, the shadows will be right under the trees.

The speed of light

Light travels fast . . . very fast. In a **vacuum,** it travels 186,420 miles (300,000 kilometers) every second. Light coming from close-by objects gets to you almost in an instant, so you see things as they happen. But even at such a high speed, light from very distant objects can take a long time to arrive. For example, light from the sun takes eight minutes to reach the earth. This means that where the sun appears in the sky is actually where it was eight minutes ago! The speed of light is an important number in **astrophysics** and, according to Albert Einstein's theory of relativity, nothing can travel faster than the speed of light.

Astronomers measure the vast distances in space in light years. One light year is about 5.9 trillion miles (9.5 trillion kilometers). These **quasars** are more than 13,000 million light years away.

Light for Seeing

We see things because light rays coming from them enter our eyes. A layer of special cells called the retina is at the back of the eye. The retina detects light and sends signals to the brain. The picture you see is built inside your brain. A **lens** is at the front of the eye, and it organizes the light to make a small picture of the scene in front of the eye on the retina.

Light and materials

We see sources of light such as light bulbs because light rays from them go straight into our eyes. We see objects that are not sources of light because light rays from other sources hit them and then go into our eyes. A simple example of this difference can be seen in the night sky. We can see stars because they make light, and planets and moons because light rays from the sun hit them. Some of these rays bounce off toward us.

When a light ray hits an object, it interacts with the object in one of three different ways. It either bounces off the object, which is called **reflection**, or it disappears into the object, which is called **absorption**, or it goes through the object, which is called **transmission**. For example, white paper reflects the light that hits it, plain glass transmits the light that hits it, and black paper absorbs the light that hits it.

Materials that do not allow any light rays to pass through them, such as wood, are called **opaque** materials. Materials that let all light rays pass through them, such as glass, are called **transparent** materials. Materials that let some light rays through but scatter them around, such as tracing paper, are called **translucent** materials.

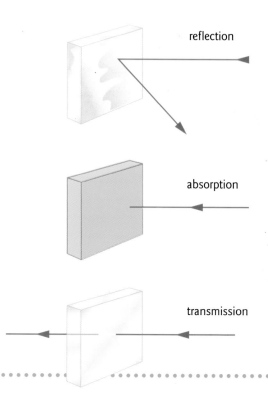

reflection

absorption

transmission

▶ Examples of light being reflected (top), absorbed (middle), and transmitted (bottom) by different materials.

Animal lights

There is a vast, dark world that little light can reach in the depths of the ocean. Many creatures that live there produce their own light called **bioluminescence**. Some creatures make light through **chemical reactions** inside special light organs in their bodies. Others have light organs that contain glowing bacteria. They use their light to send signals—either to find a mate, frighten off an attacker, attract their prey, or camouflage themselves.

Some creatures on land also glow with their own light. The best known of these are glowworms (fireflies), which are a type of beetle. They flash their lights in the darkness to attract a mate. Each species has its own pattern of flashes so that members of the same species can recognize each other's signals.

▼ Many deep-sea fish produce their own light to help them survive in the depths of the sea where no natural light can reach. This deep-sea viperfish has rows of special luminous structures along the length of its body. These are called photophores.

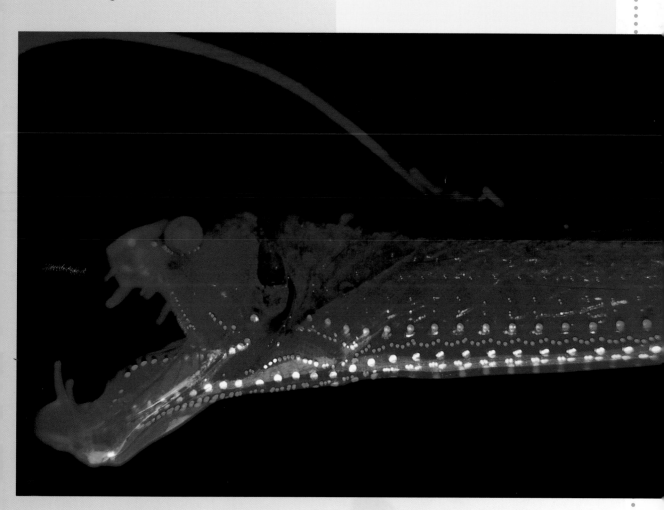

Reflection and Mirrors

Most objects are visible to us because light reflects off of them and into our eyes. Different objects look different because the materials they are made of reflect light in different ways. The shape and texture of the object matter too.

Shiny and dull

Imagine many light rays—close together and parallel to each other—coming from a small light bulb. When they hit a perfectly flat surface such as a piece of polished metal, all the rays bounce off in the same direction and stay parallel. If these rays enter your eye, they look as though they have come from a source of light in the metal. So, the metal looks shiny. Now imagine the same rays hitting a surface that is not perfectly flat, such as a piece of tissue paper. The rays still bounce off, but this time they are scattered in different directions. The rays are no longer parallel, so the surface looks dull. This is called diffuse **reflection**.

Mirrors

A mirror is a perfect reflector. It reflects all the light that hits it and keeps the pattern of rays the same. Rays of light coming from an object that hit the mirror are reflected and continue going as though they had come from an object behind the mirror. In fact, this makes the object appear to be behind the mirror. The picture of the real object that you see in a mirror is called its image. The only difference between the object and its image is that left and right appear to be reversed.

Most mirrors are made of a sheet of glass with silver paint on the back. The light reflects off the silver paint that has a very flat surface because the glass is extremely smooth.

Curved mirrors

A mirror with a curved surface makes objects look larger or smaller than they really are. A **concave** mirror has a surface that curves inward. It reflects rays toward each other. It makes them converge. A **convex** mirror has a surface that curves outward. It reflects rays away from each other. It makes them diverge. Curved mirrors are used widely in optical instruments such as telescopes.

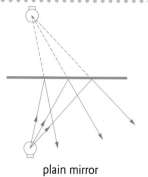

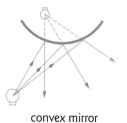

| plain mirror | convex mirror | concave mirror |

In a plain mirror, objects seem to be behind the mirror. In a convex mirror, objects seem the same but they appear smaller. In a concave mirror, objects seem to be in front of the mirror.

Super-accurate mirrors

If you look closely in a mirror, you will be able to see faint "ghost" images about one-eighth inch (3 millimeters) away from the normal image. These occur because some light reflects from the front face of the glass. This does not matter in a bathroom mirror. But in a reflecting **telescope** mirror, extra images of stars can be confusing. Reflecting telescopes have a large concave mirror made from glass. To solve the problem, the front face is covered with a super-thin coating of metal such as aluminum, which forms the reflecting surface.

The larger a telescope mirror is, the better, because it collects more light and gives a brighter image of the stars. The largest mirror ever made is almost 20 feet (6 meters) across. Telescope mirrors are made from special glass that will not expand or contract if the temperature changes. Making these huge, curved mirrors accurately is not easy because they become very heavy. One solution to this problem is to use several small telescopes instead of one large one and then combine the images from them on a computer.

This huge telescope mirror is being polished after its surface has been coated with metal.

Bending Light

When a light ray crosses the **boundary** from one **transparent** substance to another, its direction changes. This effect is called **refraction**. For example, when a light ray from an underwater object comes up through the surface and into the air, it gets bent. The only time there is no refraction is if the light ray hits the boundary at a right angle.

SCIENCE ESSENTIALS

Light rays change direction when they cross the boundary between one substance and another. This is called refraction.
A **lens** bends light in an organized way.

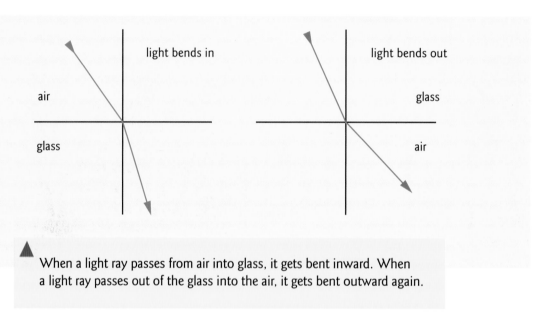

light bends in

air

glass

light bends out

glass

air

When a light ray passes from air into glass, it gets bent inward. When a light ray passes out of the glass into the air, it gets bent outward again.

Apparent depth

Refraction can make an object appear to be in a different place than where it really is. For example, when you look down into water, objects under the water appear to be higher up than they really are. The light rays from the objects are bent toward you as they leave the water.

Lenses

A lens is a specially shaped piece of glass or plastic that bends light in an organized way. The type of lens you are most likely to see is called a **convex** lens. It is the type of lens found in a normal magnifying glass. It has surfaces that curve outward, so the lens is thicker in the middle than at the edges. A **concave** lens is the opposite shape. It has surfaces that curve inward, so the lens is thinner in the middle than at the edges.

Making images

A convex lens bends light rays toward each other, and so it is often called a converging lens. Two parallel light rays passing into one side of the lens are bent so that they meet on the other side at a place called the **focal point**.

A concave lens bends rays of light away from each other, and so it is often called a diverging lens. Two parallel light rays going into one side of the lens are bent so that they appear to come from the focal point on that side.

Total internal reflection

Imagine a ray of light traveling inside a block of glass. Let's look at what happens when it reaches the surface of the glass.
• If the ray hits the surface at a right angle, it goes straight through without changing direction.
• If the ray hits at a slight angle, it goes through the surface but gets refracted.
• If the ray hits at a large angle, it does not go through at all and it is reflected. The surface of the glass acts like a mirror. This effect is called total internal **reflection**.

Total internal reflection is very useful because it does not create ghostly double images like a normal mirror. Optical instruments such as binoculars and **periscopes** use prisms instead of mirrors to turn light rays around corners. Bicycle reflectors reflect light by total internal reflection too.

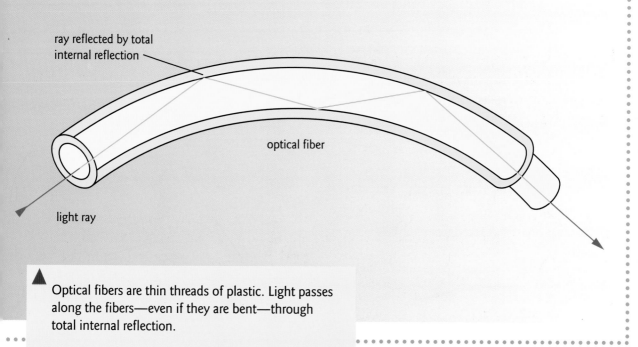

ray reflected by total internal reflection

optical fiber

light ray

▲ Optical fibers are thin threads of plastic. Light passes along the fibers—even if they are bent—through total internal reflection.

Optical Instruments

Optical instruments are devices that use **mirrors** and **lenses** to make images of objects. Examples of optical instruments are cameras, **telescopes**, and **microscopes**. Some optical instruments are simply used to capture light. Others enable us to see objects that we cannot see with the naked eye.

Capturing an image

The job of a camera is to record a scene. To achieve this, a camera points at a scene and uses a lens to create that image on film. The lens collects light rays from the scene and bends them to make an image inside the camera. The lens in the front of your eye makes an image on the retina at the back of your eye in the same way.

When the camera focuses—either automatically or manually—the distance between the lens and the film is adjusted so that the image falls accurately on the surface of the film at the back of the camera. The film is able to record the image because it contains chemicals that change when the light of the image hits them. When the film is processed, the image appears.

Seeing small things

A microscope allows you to see very tiny objects in detail. You cannot see the details with the naked eye because the lens in your eye cannot bend the light enough to focus an image onto your retina. A magnifying glass is the simplest type of microscope. Its **convex** lens bends light inward and allows you to move your eye closer to the object you are examining. Most microscopes have two lenses. The first lens, called the objective lens, makes an image inside the microscope. The second lens, called the eye lens, magnifies that image.

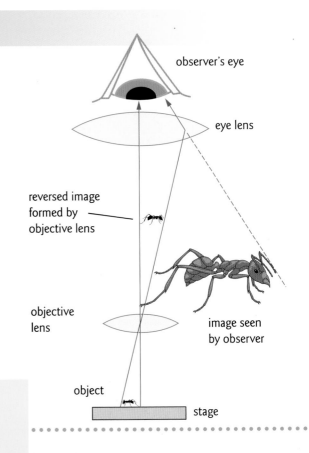

observer's eye

eye lens

reversed image formed by objective lens

objective lens

image seen by observer

object

stage

▶ A microscope bends light rays coming from an object so that the object appears to be much bigger than it is.

Seeing distant things

Telescopes make distant objects look larger. The simplest type of telescope is the refracting telescope, which has two lenses. The first lens, called the objective lens, makes an image of the distant object. The second lens, called the eye lens, magnifies that image. A reflecting telescope has a large **concave** mirror instead of an objective lens.

Digital images

Digital cameras take photographs electronically. Instead of film in the back of the camera, there is a special **microchip** called a charge-coupled device (CCD) that records the image. The face of the CCD is covered by a grid of thousands of tiny squares called pixels. Each square is charged with electricity before the image made by the camera is allowed to fall on the grid. The color and brightness of the light of the image affect the charge in the pixels. Electronic circuits detect the remaining charge and record the color and brightness of the image at each pixel. This information can be loaded into a computer where the images can be viewed, manipulated, and printed.

The CCD in an average compact camera has a grid of a few hundred pixels wide, with a few hundred thousand pixels in all. Telescopes and microscopes use CCDs too. They are much larger and more detailed than those in cameras and have a total of tens of millions of pixels. The images from these telescopes and microscopes can be analyzed and **enhanced** by computers.

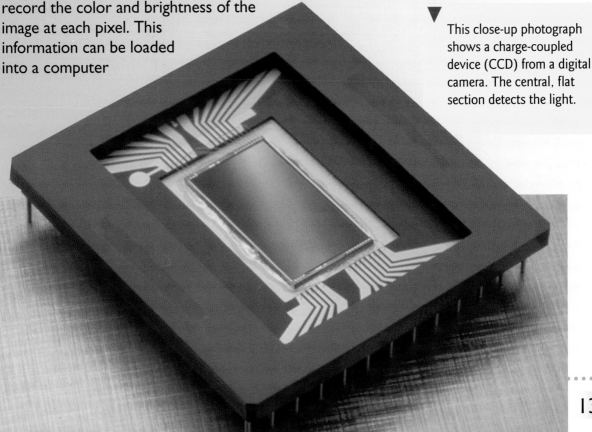

▼
This close-up photograph shows a charge-coupled device (CCD) from a digital camera. The central, flat section detects the light.

Color

SCIENCE ESSENTIALS

White light is a mixture of different colors of light.

The range of different colors is called the **color spectrum**.

A **filter** only allows the light of its own color through.

The light that comes from light bulbs or the sun looks white and is called **white light**. However, white light does not really exist. In fact, it is a mixture of equal amounts of many different colors of light that combine together to look white.

Rainbow colors

Sometimes white light gets split into its different colors, which makes the colors visible. This happens because the different colors are refracted by slightly different amounts when they cross a **boundary**. For example, when white light crosses a boundary, the red part is bent more than the blue part. This effect is called **dispersion**. This is what happens when a rainbow forms.

Sunlight is dispersed as it is refracted around rain drops.

The colors in a rainbow are called the colors of the spectrum. They are red, orange, yellow, green, blue, indigo, and violet. The colors do not change abruptly from one to the next, but they blend together smoothly.

How objects have color

You already know that when light hits an object, it is either reflected, transmitted, or absorbed. But why does one object look different from another? The answer is that the objects affect the light that hits them in different ways. For example, a green apple looks green because it reflects only green light. It absorbs all the other colors of light. This means that only green light is visible when you look at the apple.

▶ When white light hits this apple, all of the colors of the spectrum are absorbed except green, which is reflected. This makes the apple look green.

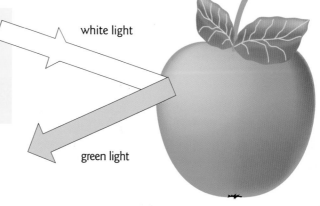

white light

green light

green apple

Filters

A filter is a piece of glass or plastic that only allows some colors of light through. It absorbs all the other colors. For example, a red filter only allows red light to pass through, and it stops all of the other colors. So when you shine a white light through a red filter, only red light comes out on the other side.

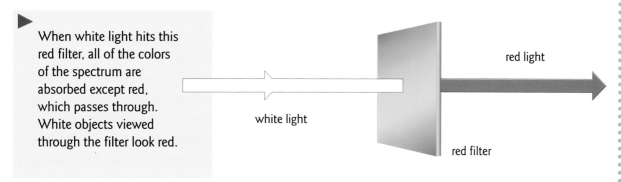

▶ When white light hits this red filter, all of the colors of the spectrum are absorbed except red, which passes through. White objects viewed through the filter look red.

white light

red light

red filter

Flames and spectroscopy

When any substance is heated to a very high temperature, it begins to emit light. The color of the light can help to determine what the substance is. One simple test in chemical analysis is the flame test, which is used to find what sort of metal there is in a substance. For example, a substance that contains copper burns with a green-blue flame, and a substance that contains sodium burns with an orange flame.

In fact, substances give out a mixture of different colors of light when they are heated. In a flame test, these are all mixed together. To identify the colors, a device called a spectroscope is used. This disperses the light from the substance, showing which colors of the whole spectrum are present in the light and in what proportions. This is called an **emission spectrum**. Every chemical **element** has its own characteristic emission spectrum. This means that a spectroscope can be used to find out what elements there are in a mixture.

▶ Each tiny spectrum is made by splitting light coming from an object in space. Different objects (stars, **quasars**, and so on) create a different spectrum.

Primary Colors

Although there is a range of colors in the **color spectrum**, there are some special colors that can be combined together to create any color. These colors are called **primary colors**. There are three primary colors of light.

SCIENCE ESSENTIALS
The primary colors of light are red, green, and blue.
Any colors can be made by adding the primary colors in different proportions.

Primary colors of light

The primary colors of light are red, green, and blue. By mixing these colors in different proportions, any color of the spectrum can be made. This process is called color addition because new colors are made by adding different colors of light together. Mixing red, green, and blue light in equal proportions makes **white light**. An example of color addition is in television. The screen emits only red, green, and blue light, but it adjusts the brightness of each one on different parts of the screen to create the colored picture.

Colors of paint

The colors in inks and paints are created by **pigments**. You see these colors because of **reflection**. When white light hits an object, the pigments absorb some colors and reflect others. This is called color subtraction. The primary colors of pigments are yellow, cyan, and magenta. When they are all mixed together, all the colors of the spectrum are absorbed and you see black.

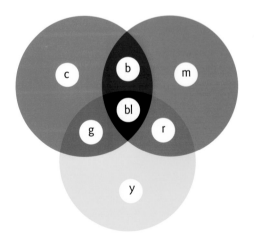

Adding the primary colors of light (left) and pigments (right).

r = red; g = green; b = blue
c = cyan; m = magenta; y = yellow
bl = black; w = white

Colors in colored light

In white light, colored objects look the color they are because they reflect certain parts of the white light. If they are illuminated by colored light, they look a different color. For example, a green apple illuminated by a red light looks black because there is no green light to be reflected and the red light is absorbed.

Printing in color

If you look at the color photographs in this book with a magnifying glass, you will see hundreds of tiny dots of color. But there are only four colors—cyan, magenta, yellow, and black, or CMYK for short. The different colors in the pictures are produced by adjusting the size of the dots. From a distance, the dots blend together to appear as colors. Black ink is used because the other three make brown, not black, unless you use quite a lot of each. Black is also normally used for the text.

When the page is printed, the four colors are added one by one, using the four colored inks. Gradually the pictures are built up. A book like this is printed on a four-color printing press. This printing process prints color books and magazines with just four colors of ink. Color ink-jet printers work in the same way by firing tiny blobs of colored ink onto the paper.

However, CMYK printing cannot reproduce certain colors, so sometimes special colors are added to the basic four. This is often the case for printing fine art prints, which need a subtle range of colors. Metallic colors, such as gold and silver, or **fluorescent** colors, are added with separate inks too.

▶ A magnified section of a color picture from an ink-jet printer. You can see the tiny dots of cyan, magenta, yellow, and black ink, which are normally too small to see.

Using Color

Try spending a day looking out for color. Look at the decorations in your home, your clothes, colors of plants as you travel to school, signs around your school, and books and the pictures and lettering in them. Try to imagine why all the objects you see are the colors they are.

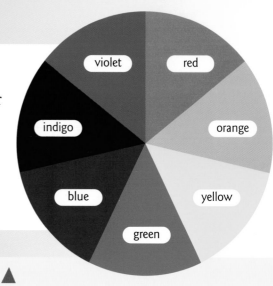

A sample color wheel using only seven colors.

Colors for decoration

Interior decorations, clothes, books, and magazines use color to look good. The trick to creating a pleasing color scheme is to choose the right color combination. Get it wrong and the result is almost impossible to look at!

If you think about grabbing the two ends of the **color spectrum** and pulling them together so that red is next to violet, you get a color wheel. Mixing colors that are opposite to each other on the wheel, such as red and green or blue and orange, makes a contrasting, vibrant color scheme.

Mixing colors that are close to each other on the wheel, such as blue and indigo, creates a harmonious color scheme.

As you move along the spectrum from red to violet, the colors change from "warm" to "cool." Decorating with warm colors, such as reds and oranges, makes a room feel cozy. Decorating with cool colors, such as greens and blues, makes a room feel cool and airy.

Getting the message

Colors are also good at getting messages across quickly. Flashes of color in a bland background catch the eye and attract people's attention. Signs are more likely to be read if they are in red type or have a red border around them. Certain colors have special meanings. For example, red is often used for danger, yellow is for warnings, and green normally stands for "okay." You can see these colors at work in traffic signals.

Bright red is a color that is not often seen in nature or used in decorating, so a bright red sign stands out from the background.

Colors in nature

Animals and plants use color to their advantage too. Many flowering plants have brightly colored petals that advertise to birds and insects that there is nectar inside the flowers. As the animals visit flowers, they carry pollen from one to the other, which helps the plant to reproduce.

In many bird species, the male has brightly colored feathers or other body parts for display during courtship in order to attract a mate. Some animals use color for camouflage, so that they can hide from predators among leaves or grass. Other animals such as ladybugs, snakes, and frogs do quite the opposite. They have bright spots or stripes of red or yellow on their skins to warn predators that they are poisonous or bad-tasting.

Perhaps the most amazing are those animals that can change color at will. Lizardlike chameleons change the color of their skins to match their background, so they can camouflage themselves almost anywhere. Cuttlefish have special droplets of different-colored **pigments** in their skin. By changing the sizes of the droplets, they change their colors. Cuttlefish are believed to send messages to each other by making ripples of color move along their bodies.

▼
The bright colors of these poison arrow frogs warn predators that there are chemicals in their skin that are deadly poisons.

19

Invisible Light

Light is a type of **electromagnetic radiation**. It is made up of changing electric and magnetic fields. Physicists often think of this radiation as traveling in **waves**, just as waves travel across water and spread out from where they are made. Light is not the only type of electromagnetic radiation. It is just one member of a whole family of electromagnetic waves called the **electromagnetic spectrum**.

SCIENCE ESSENTIALS

Light is a type of electromagnetic radiation or wave.
Light is part of a family of waves called the electromagnetic spectrum.

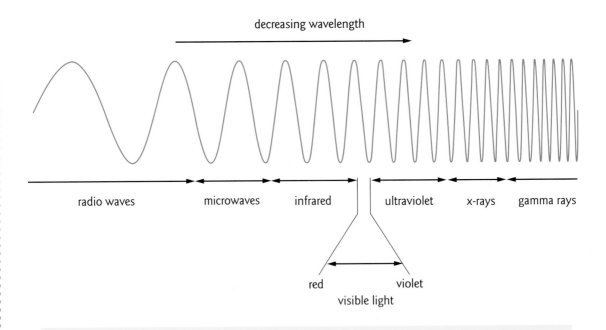

decreasing wavelength

radio waves   microwaves   infrared   ultraviolet   x-rays   gamma rays

red   violet
visible light

A simplified diagram of the electromagnetic spectrum. In reality, the wavelengths at the left end are millions and millions of times greater than those at the right end.

The electromagnetic spectrum includes waves that seem very different from light, such as radio waves, microwaves, and x-rays. But all of these waves travel at the speed of light and carry **energy** away from where they are made. Moving from one end of the electromagnetic spectrum to the other, the **wavelength** of the waves decreases and the **frequency** increases.

The part of the **color spectrum** that makes up the light we see is called the visible spectrum. Each color is made up of electromagnetic waves of a different wavelength. We see the different colors because our eyes can detect the different wavelengths.

Infrared and ultraviolet

Sunlight contains two more types of radiation that are often included in the color spectrum even though they are invisible. **Infrared radiation** is below red light, and **ultraviolet radiation** is above violet.

Infrared radiation is a form of heat energy. It is given off by all objects, but it gets stronger as the object gets hotter. You can feel infrared radiation as warmth on your skin.

Ultraviolet (UV) radiation is the type of radiation that tans fair skin in the sun. Small amounts of UV are good for you, but large amounts can cause sunburn and even skin cancer. Some security markings are made with special ink that only shows up when UV light is shined on it. If items that are marked with the ink are stolen and then recovered, the police can use a UV lamp to identify the items.

The invisible sky

Taking photographs of space can tell us a certain amount about the stars, **galaxies**, and other objects in our universe, such as their positions. However, much more information can be found by collecting the nonvisible types of radiation coming from space such as radio waves and x-rays. Visible light coming from distant stars is often blocked by huge clouds of dust and gas far out in space. But other types of radiation from the stars can pass through the clouds. Looking for this radiation can make the stars and galaxies "visible."

Infrared astronomy shows the heat coming from objects in space. The most intense heat comes from the center of our galaxy where many stars are being formed. X-ray astronomy shows that x-rays are coming from all over the sky.

Most of them come from hot gas in our galaxy. Radio astronomy reveals many things including sweeping beams of radio waves coming from tiny dead stars called pulsars.

▶ Infrared radiation is also used to photograph the earth from space. This image shows the heat coming from part of Japan. Plants look red, buildings look blue, and water looks black.

Waves and Sound

A **wave** happens when a movement is repeated again and again. The waves we see most often are the waves that move across the surface of an ocean or lake. However, light and sound move as waves too, even though we cannot see the waves in action. All waves carry **energy** away from where they are made.

SCIENCE ESSENTIALS

Waves happen when movements are repeated again and again.
The **amplitude** is the height of a wave.
The **wavelength** is the distance between one wave **crest** and the next.
The **frequency** is the number of wavelengths that pass a point every second.
Sound is made by vibrating objects.
Sound spreads through the air in waves.

Moving in waves

As an ocean wave travels past a point in the water, the particles of the water move up and down. As they move, they make the particles next to them move too. So the wave moves through the water, even though the particles themselves do not actually move along with the wave. Because the water particles in waves actually move at right angles to the direction of the waves, water waves are called transverse waves.

The other type of wave is called a longitudinal wave. In a longitudinal wave, the particles move backward and forward in the same direction as the waves are traveling. A wave going down a long, slinky spring is an example of a longitudinal wave. Pulling briefly on one end of the spring pulls the coils at that end apart. Each coil then moves one way and then the other as the wave moves along the spring.

Wave words

The amplitude of a wave is the height of the wave. The larger the amplitude, the bigger the movements that make the wave. The wavelength is the distance between one wave crest and the next. The frequency is the number of crests that pass a point every second. Frequency is measured in hertz (Hz).

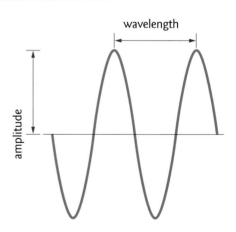

How amplitude, wavelength, and frequency are measured.

Light and sound as waves

Light can be thought of as waves of **electromagnetic radiation**. Sound travels as waves too. Sound is caused by vibrations. Imagine "twanging" a plastic ruler on the edge of a table. The ruler vibrates up and down very quickly. These vibrations make the particles of air next to the ruler vibrate backward and forward. These movements are passed on to the next particles, and so waves of vibrations spread. This is how sound travels through the air. As a sound wave passes a point in the air, the particles of the air are pulled apart and then squashed together.

▲ Sound waves spread in all directions from a source of sound.

*W*ave shapes

Although you cannot see a sound wave, you can draw it as a line graph. The line looks like a wave on water. It shows how the particles in the air are squeezed and stretched as time passes by. A very pure sound such as the sound from a tuning fork has a smooth wave shape like the one in the diagram on the opposite page. A violin makes a much more jagged sound wave, but the same shape is repeated again and again. Noise, such as the sound of traffic, has a random wave shape. Wave shapes can be seen by connecting a microphone to an **oscilloscope**.

▼ Sections of sound waves taken from a rock music CD. They are a mixture of sounds from different musical instruments.

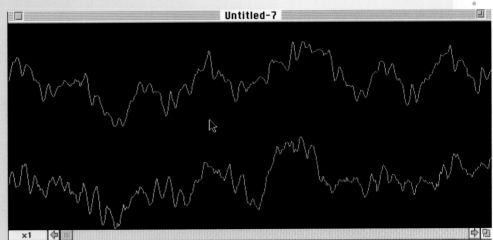

Traveling Sound

Any object that vibrates—from leaves moving in the wind to the cone of a **loudspeaker**—causes sound. Sound travels through solids and liquids as well as the air. But unlike light, it cannot travel through a **vacuum** because it can only travel by using the particles in a substance.

SCIENCE ESSENTIALS

Sound travels in a material. It cannot travel in a vacuum.
Sound travels away from its source.
Sound travels at different speeds in different materials.
An **echo** happens when a sound bounces off a surface.

Echoes

Just as waves in water bounce when they hit a surface like a harbor wall, so sound waves bounce (are reflected) when they hit surfaces. Smooth, hard surfaces such as bare walls and cliffs are best at reflecting sound. If you hear a sound directly from its source and it reflects off a surface so that you hear it again, you are hearing an echo. Not all surfaces are good at reflecting sound. Soft surfaces such as curtains absorb sound instead. This is why an empty room with bare walls echoes, but a furnished room does not echo.

Echoes are important in a concert hall, so that the music sounds rich but without much echo. The special mushroom-shaped roof in this hall bounces sound back down to the audience.

Searching with sound

When your voice echoes, you know a surface nearby has reflected the sound. Some animals such as dolphins and bats create sound and use the echoes to detect objects around them. This is called **echolocation**. The time the sound takes to return indicates the object's distance. Many ships have a device that uses echolocation to detect underwater objects and to measure the water's depth.

Sound waves are also used to find underground objects such as rocks that might contain oil and gas. Large machines fire strong sound waves into the ground, and the **reflections** are detected by sensitive microphones. Sound reflects off the **boundaries** between different types of rock. Expert **geophysicists** examine the results and try to determine the structure of the rocks.

The speed of sound

Sound travels very fast, but not as fast as light. You can check this during a thunderstorm. You will see the flash of lightning before you hear the loud noise created by the lightning. In the air at sea level, sound travels at about 372 yards (340 meters) per second. It slows down higher in the atmosphere because the air is much thinner. Just 10,900 yards (9962 meters) higher, above the summit of Mount Everest, sound travels at 328 yards (300 meters) per second.

Sound travels much faster in liquids and solids than it does in the air. The speed of sound in water is about 1640 yards (1500 meters) per second, and in glass it is 5470 yards (5000 meters) per second.

Some objects, such as jet aircraft and bullets, can travel faster than the speed of sound. Their speed is normally measured by Mach numbers. Mach 1 is equal to one times the speed of sound; Mach 2 is equal to twice the speed of sound, and so on. Faster speeds than the speed of sound are called supersonic speeds. An object traveling at a supersonic speed creates a shock wave of sound, which is heard as a loud bang called a sonic boom.

▼ Humpback whales are famous for the various groaning, chirping, and crying noises they make to communicate. The sound travels well underwater and can be heard more than 90 miles (144 kilometers) away.

Hearing Sound

Our ears are used for detecting sound. A thin membrane called the eardrum is just inside your ear. When a sound wave enters your ear, the changing pressure makes the eardrum vibrate in and out. In turn, this makes tiny bones attached to the eardrum vibrate too. The bones pass the vibrations to the inner ear where the vibrations make tiny hairs move. These movements are detected by **nerve cells** that pass messages to your brain.

Loudness and pitch

The larger the amplitude of a sound, the louder it sounds. The vibrations that cause most of the sounds we hear such as people talking are too small to see. But some vibrations such as very loud drum beats are so large you can actually feel the sound waves as well as hear them.

The higher the frequency of a sound, the more high-pitched it sounds. You can probably hear sounds with frequencies between about 20 hertz and 20,000 hertz (20 kilohertz). But when you grow older, the range of frequencies you can hear will probably be reduced. Sound with an extremely high frequency is called ultrasound. We cannot detect it, but it has many applications in science, engineering, and medicine. For example, an ultrasound scanner is used to search for potentially dangerous cracks in the metal of aircraft.

Animal hearing

The range of frequencies that animals can detect is different from the range that humans can hear. For example, cats and dogs can hear sound with higher frequencies, and elephants can hear sound with lower frequencies. Bats can hear frequencies as high as 120,000 hertz, which allows them to detect the echoes of the high-pitched squeaks they use for **echolocation**. Many animals also have large ear flaps that allow them to hear very quiet sounds.

Dangers of sound

The loudness of a sound is measured in units called decibels (dB). On the decibel scale, an increase of 10 decibels represents a ten-times increase in loudness. This means that a sound measuring 50 decibels is ten times as loud as a sound measuring 40 decibels. A sound of 60 decibels is 100 times as loud. Rustling leaves measure about 20 decibels; when you speak normally to your friends your speech measures about 50 decibels; and loud music measures about 100 decibels.

Sounds with a loudness of more than 120 decibels can damage your ears because the vibrations of the air are so great. Sounds this loud are caused by explosions or rockets taking off. Quieter sounds can also damage your ears if you listen to them long enough.

Sounds can be annoying as well as dangerous. Most of the time, unless we live in the country, we can hear background noise such as cars on the road, aircraft flying past, or music playing next door. This is called noise pollution. People who live near busy roads or airports may receive grants to help them to soundproof their homes. Noise is often the cause of disputes between neighbors.

► A jet airplane taking off creates a loud roar with its engines on full power. Older airplanes from the 1960s and 1970s have extremely noisy engines. Many airports in urban areas have a ban on night takeoffs.

Musical Sounds

At its simplest, music consists of a series of sounds called tones played one after the other. In music, each tone has a certain **frequency.** Playing certain tones after, or with, others makes a pleasing sound. Playing tones of other frequencies, or some tones after others, creates unpleasant sound. All musical instruments work by making the air vibrate in some way. There are three main families of acoustic instruments (that is, instruments that do not create sound electronically): string, wind, and percussion.

String instruments

The guitar, violin, and piano are examples of string instruments. A string instrument has a set of metal or nylon strings, each of which vibrates at its own natural frequency. The frequency depends on the length and thickness of the string and how tight it is stretched. Shorter, thinner strings with more tension make higher tones than longer, fatter strings with less tension. To make the string vibrate, it is plucked (as in a guitar), hit with a hammer (as in a piano), or drawn with a bow (as in a violin). A piano has a separate string for each tone. To change the tone made by a string on a guitar or violin, the player shortens the string by pressing a finger onto it.

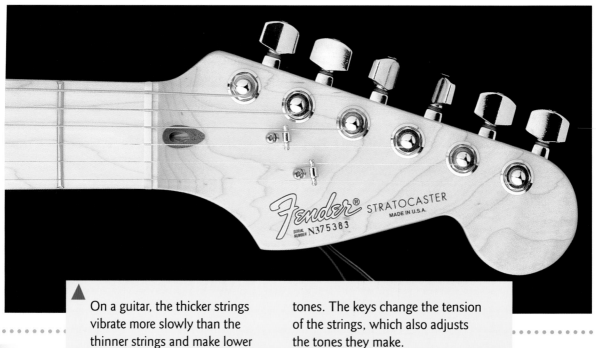

On a guitar, the thicker strings vibrate more slowly than the thinner strings and make lower tones. The keys change the tension of the strings, which also adjusts the tones they make.

Wind instruments

The trumpet, flute, and clarinet are examples of wind instruments. All wind instruments have a long tube of air inside of them, and the sound they make is caused by the air in the tube vibrating. The frequency of the tone that the instrument plays depends on the length of the tube. The longer the tube, the slower the air vibrates within it and the lower the tones that it can play.

The air is made to vibrate when the player blows air past a reed that then vibrates (as in a clarinet), or blows across a hole in the tube (as in a flute), or vibrates his or her lips against the end of the tube (as in a trumpet). In most wind instruments, the length of the tube, and so the tone the instrument plays, is changed by covering or uncovering holes in the tube, either by putting fingers over the holes or by pressing valves.

Percussion instruments

A percussion instrument makes tones when it is struck in some way, and so it is made to vibrate at its natural frequency. Simple percussion instruments such as triangles and bells only produce one tone. A xylophone has many blocks of wood or metal, each of which makes its own tone. Percussion instruments such as drums and gongs make noise rather than tones, but the pitch of the noise depends on their size.

▶ A synthesizer is an electronic musical instrument. Its complex electronic circuits can create sounds that sound almost exactly like any real instrument. They are excellent tools for composing music.

Glossary

absorption process by which light disappears into a material

amplitude height of a wave from its center to its peak

astrophysics study of the physics of objects in space such as planets, stars, and galaxies

bioluminescence light created by animals such as glowworms

boundary where one material stops and another begins

chemical energy energy stored in chemicals that can be released when the chemical takes part in a chemical reaction

chemical reaction when one or more chemicals divide or combine together to form new chemicals

color spectrum range of colors of visible light, from red to violet

concave describes a surface that dips inward in the center such as the inside of a spoon

convex describes a surface that extends outward in the center such as the underside of a spoon

crest highest point of a wave. The lowest point is called a trough.

digital having to do with numbers

dispersion process by which different colors of light are refracted by different amounts and split up white light

echo sound that is heard again after being reflected from a surface

echolocation method of finding objects by listening for the echoes that bounce off of them

electrical energy energy carried by an electric current flowing around a circuit

electromagnetic radiation energy carried from place to place by a changing magnetic and electric field

electromagnetic spectrum family of waves including light, radio waves, microwaves, and x-rays

element simplest substance that exists that cannot be broken into simpler substances

emission spectrum mixture of different colors of light that a material emits when it is heated

energy ability to make something happen

enhance to improve or make better

filter piece of colored glass or clear plastic that stops some colors of light but allows others to pass through

fluorescent describes a material that gives off light when it is hit by electromagnetic radiation

focal point point at which light rays meet after passing through a lens or reflecting off a mirror

frequency number of wave crests that pass a point every second

galaxy large group of stars

geophysicists scientists who study the physics of the earth and the earth's structure

infrared radiation invisible rays that are similar to red light rays and carry heat energy from place to place

lens specially-shaped piece of glass or clear plastic that bends light rays in an organized way

loudspeaker device that turns a changing electric current (and electrical signal) into sound

mechanical energy energy that an object has because it is moving, spinning around, or is squashed or stretched

microchip small piece of semiconductor material (normally silicon) with a complex electronic circuit built into it

microscope optical instrument used to examine objects that are too small to be seen with the naked eye

mirror smooth shiny surface that reflects all the light that hits it

nerve cell animal cell that carries signals to or from the animal's brain

nuclear energy energy stored in the nucleus (center part) of an atom, which is released when the nucleus splits up or combines with the nucleus of another atom

opaque describes a material that does not let light pass through it

oscilloscope instrument that shows how an electrical current changes by drawing a line on a screen

periscope optical device that turns light rays around two corners. This allows the user to get a view from above his or her head.

pigments natural chemicals that give paints, inks, and animals' skins their color

primary colors the colors red, green, and blue, which can be combined to create any color of the color spectrum

quasar object in deep space that is thought to be a galaxy with a high-energy center

reflection process by which light or sound bounces off of a surface

refraction process by which light rays change direction when they cross a boundary between two substances

shadow area of darkness created when an object blocks light

solar energy any energy that originates from the sun, such as sunlight

telescope optical instrument used to examine distant objects

translucent describes a material that scatters the light that passes through it

transmission process by which light passes right through a material

transparent describes a material that allows light to pass through it

ultraviolet radiation invisible rays in sunlight that are above violet in the color spectrum

vacuum space that contains nothing, not even air

wave movement of particles (or changes in electromagnetic fields) that carry energy from place to place, such as sound waves or light

wavelength distance between one crest and the next on a wave

white light light made by mixing all the colors of the color spectrum in equal proportions

More Books to Read

Friedhoffer, Robert. *Light*. Danbury, Conn.: Franklin Watts, Incorporated, 1992.

Gardner, Robert. *Experiments with Light & Mirrors*. Springfield, N.J.: Enslow Publishers, 1995.

Kerrod, Robin. *Light & Sound*. Tarrytown, N.Y.: Marshall Cavendish Corporation, 1995.

Index